SPACE EXPLORERS

A coloring book adventure

I0049078

LUNA OWL
PRESS

LET'S GO ON AN ADVENTURE THROUGH SPACE!

MERCURY IS THE SMALLEST PLANET IN OUR SOLAR SYSTEM!

It's covered in craters from ancient impacts.

INSIDE MERCURY

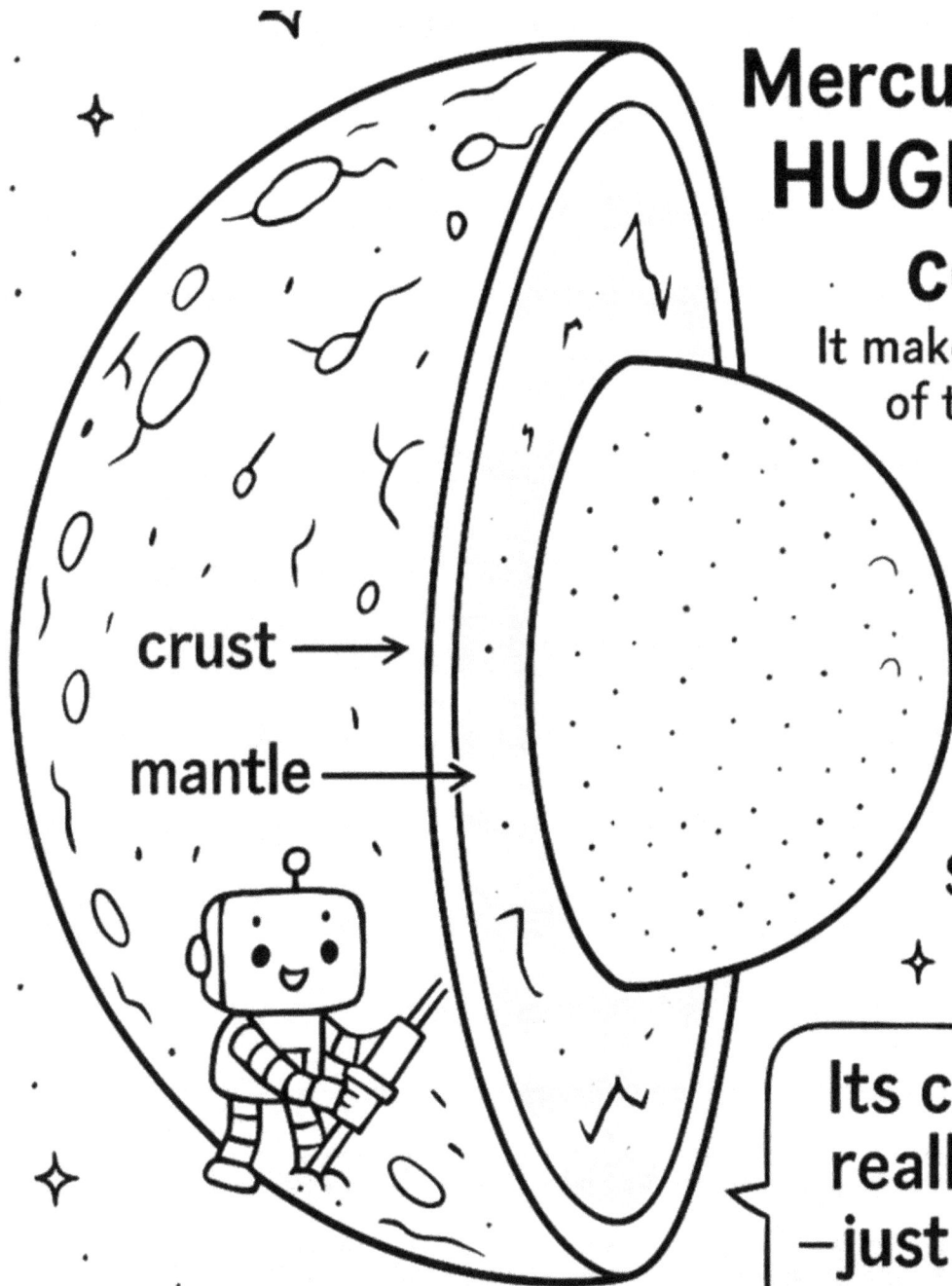

crust →

mantle →

Mercury has a HUGE metal core!

It makes up most of the planet's inside!

Super Hot!

Its crust is really thin——just a shell over the core!

VENUS

Venus is the hottest planet
in our solar system!

(Even hotter than Mercury!)

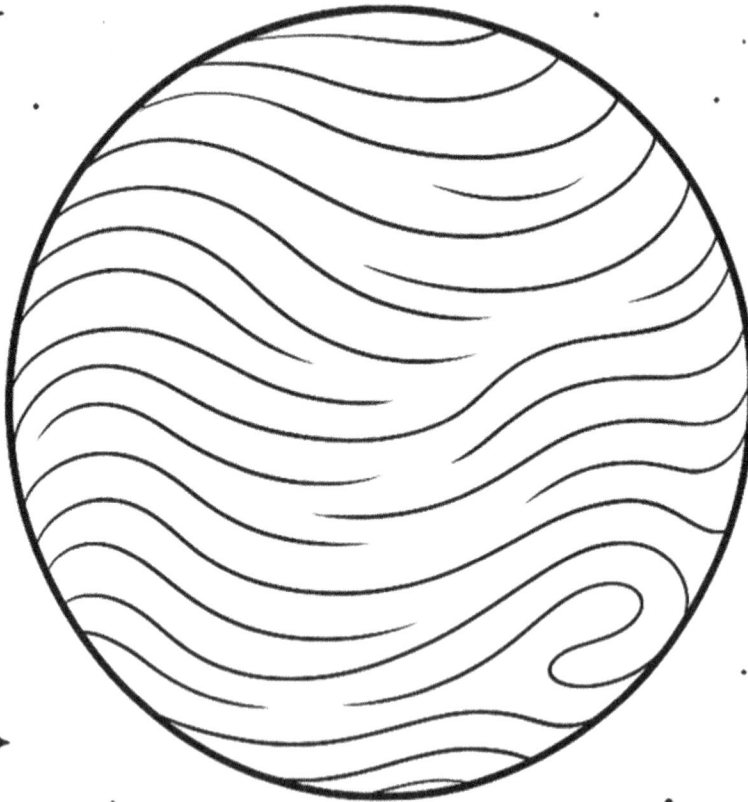

Its thick clouds trap
heat like a giant oven.

INSIDE VENUS!

CLOUDS OF SULFURIC ACID!

Venus has more volcanoes than any other planet!

Its clouds are made of acid and trap heat like a giant oven!

850°F

CRUST

MANTLE

CORE

The surface is hot enough to melt lead.

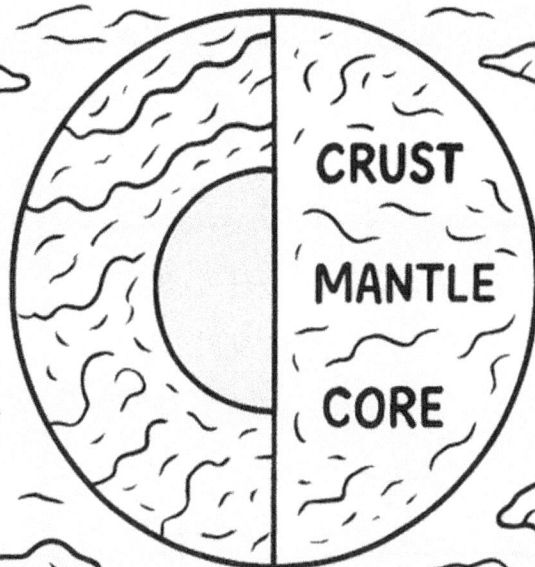

EARTH

Earth is the only planet we know with life!

About 70% of Earth's surface is covered in water!

EARTH

CRUST

OUTER CORE

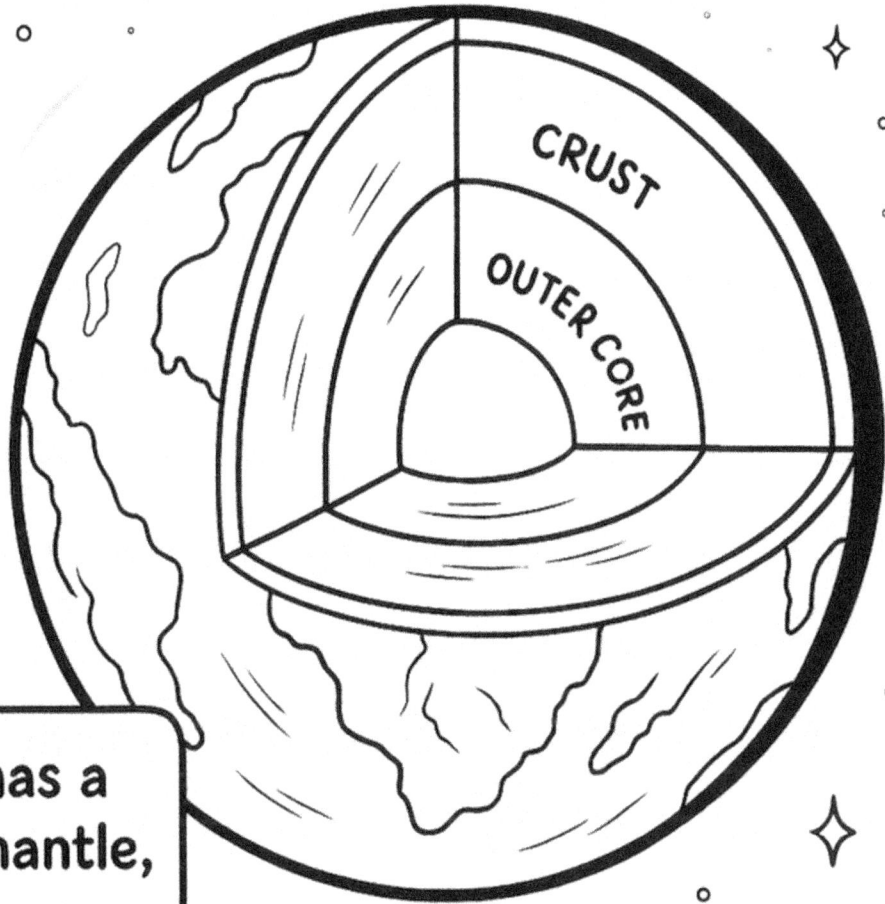

Earth has a
crust, mantle,
and core.

The inner core is as
hot as the surface
of the sun!

MOON

Sea of Tranquility

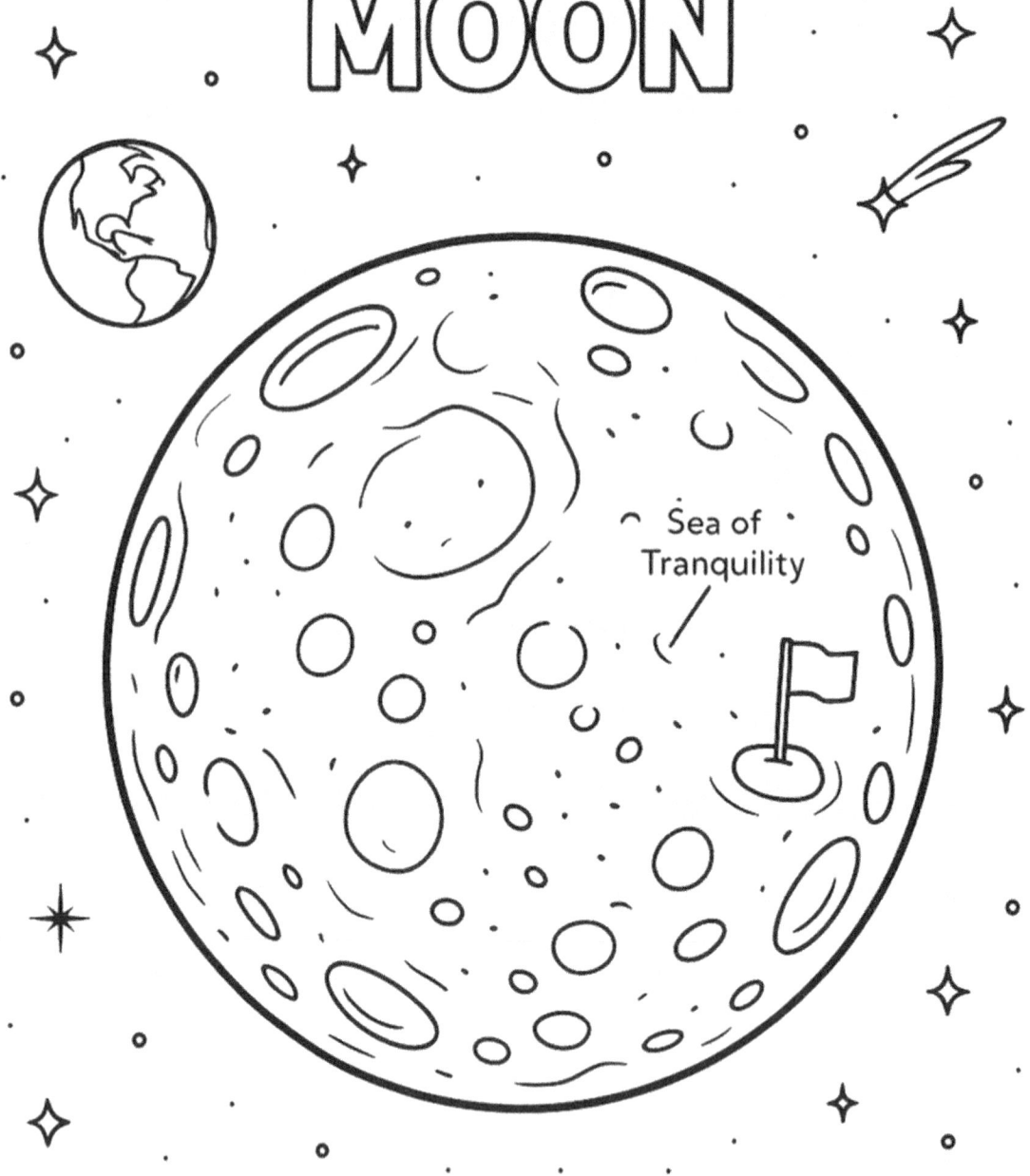

The Moon is Earth's only natural satellite!

It has craters, mountains, and flat plains called maria.

It takes about 27 days to orbit Earth.

HOW WE GOT TO MOON

In 1969, astronauts walked on the Moon for the first time!

Only 12 people have ever walked on the Moon!

MARS

Mars is called the Red Planet because of its rusty soil!

It has th tallest volcano and deepest canyon in the whole solar system!

INSIDE MARS

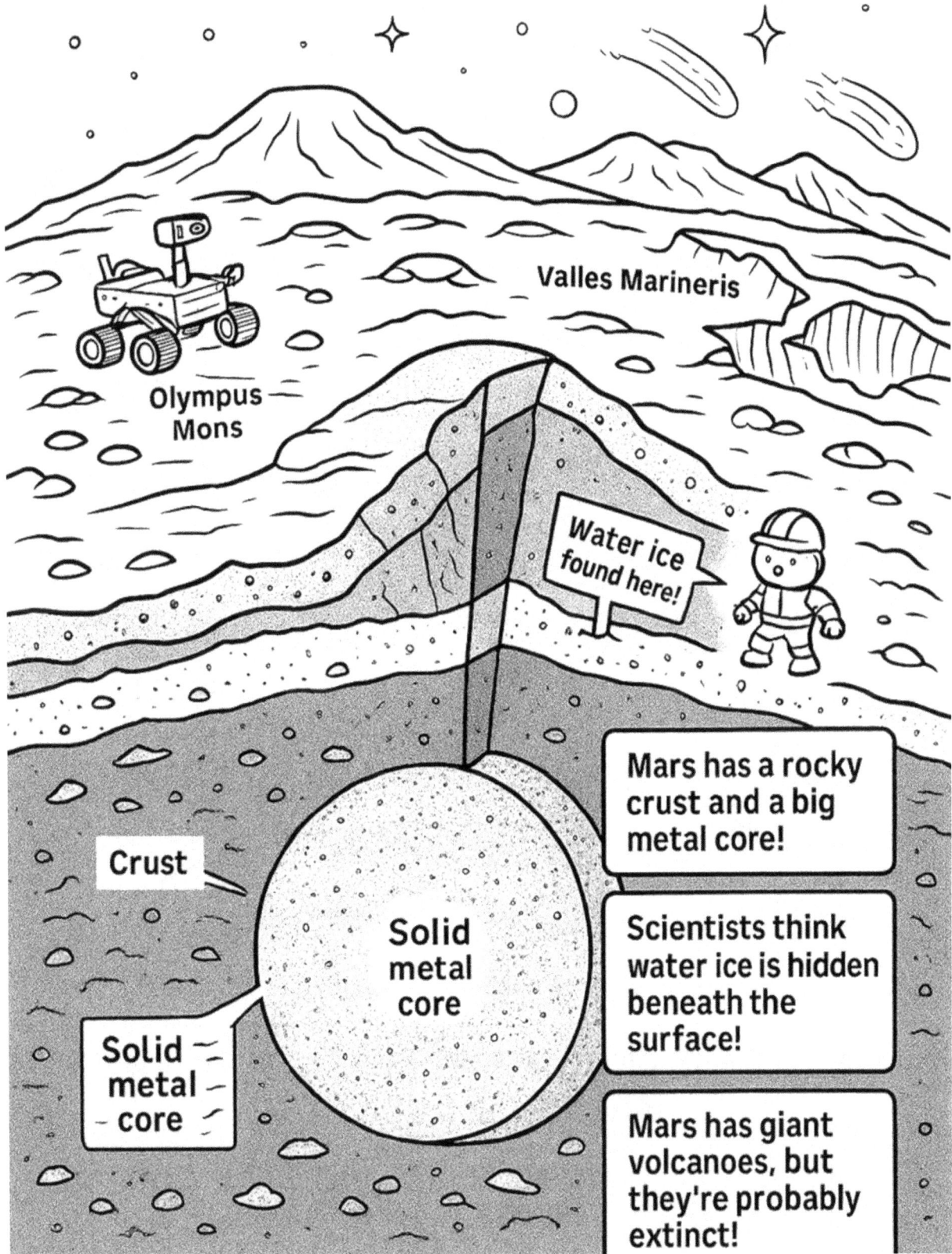

Valles Marineris

Olympus Mons

Water ice found here!

Crust

Solid metal core

Solid metal core

Mars has a rocky crust and a big metal core!

Scientists think water ice is hidden beneath the surface!

Mars has giant volcanoes, but they're probably extinct!

THE RINGED BEAUTY

SATURN IS KNOWN FOR ITS BEAUTIFUL RINGS
MADE OF ICE AND ROCK!

IT HAS MORE THAN 140 MOONS—INCLUDING TITAN,
WHICH IS BIGGER THAN MERCURY!

INSIDE SATURN

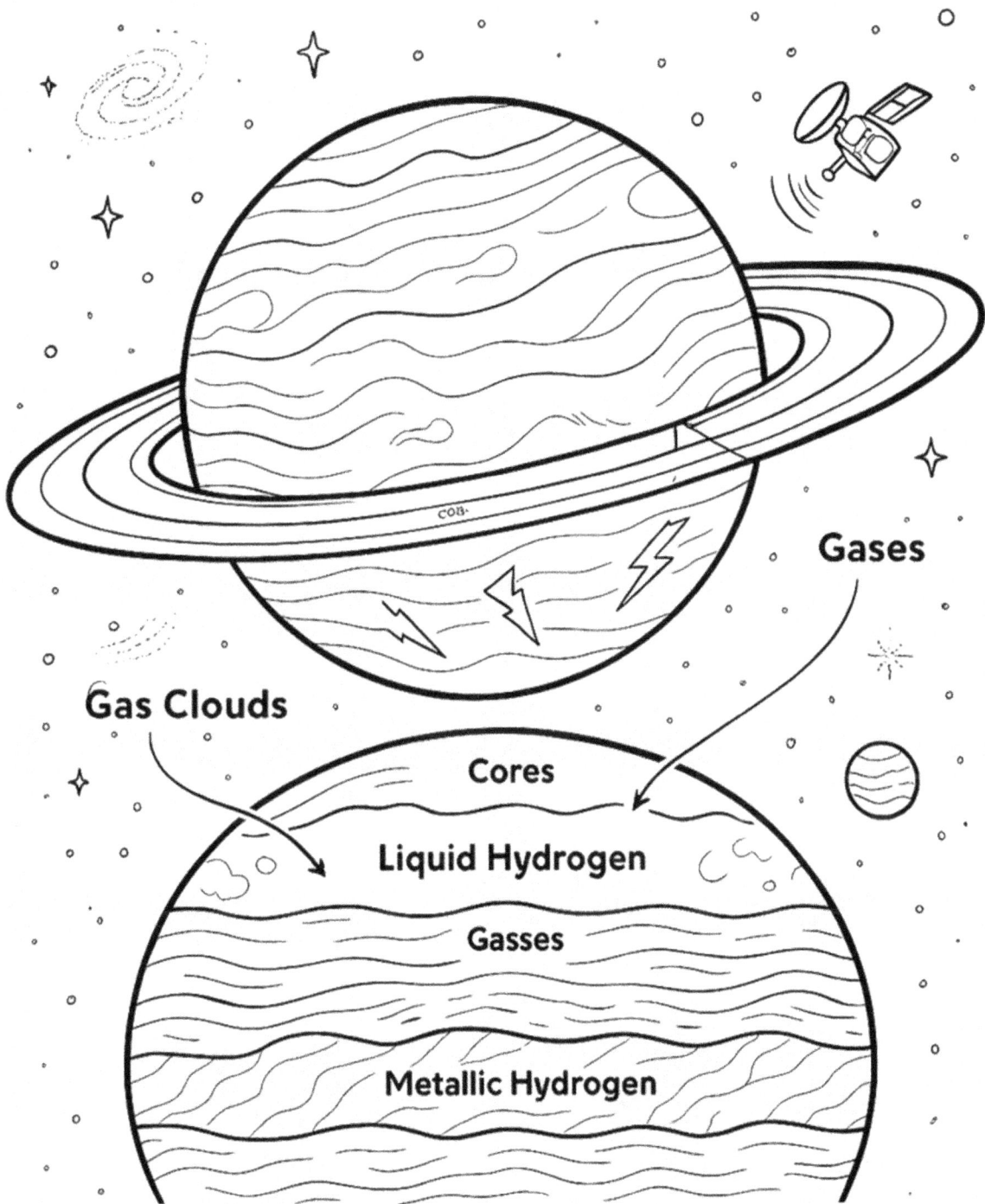

Gases

Gas Clouds

Cores

Liquid Hydrogen

Gasses

Metallic Hydrogen

Saturn has thick gas clouds in its atmosphere!

Deeper inside, the gas becomes a dense liquid

Inside Saturn is a hot, dense core

Uranus

Uranus spins on its side—it's tilted almost 98 degrees!

It has faint rings and at least 27 moons!

INSIDE URANUS!

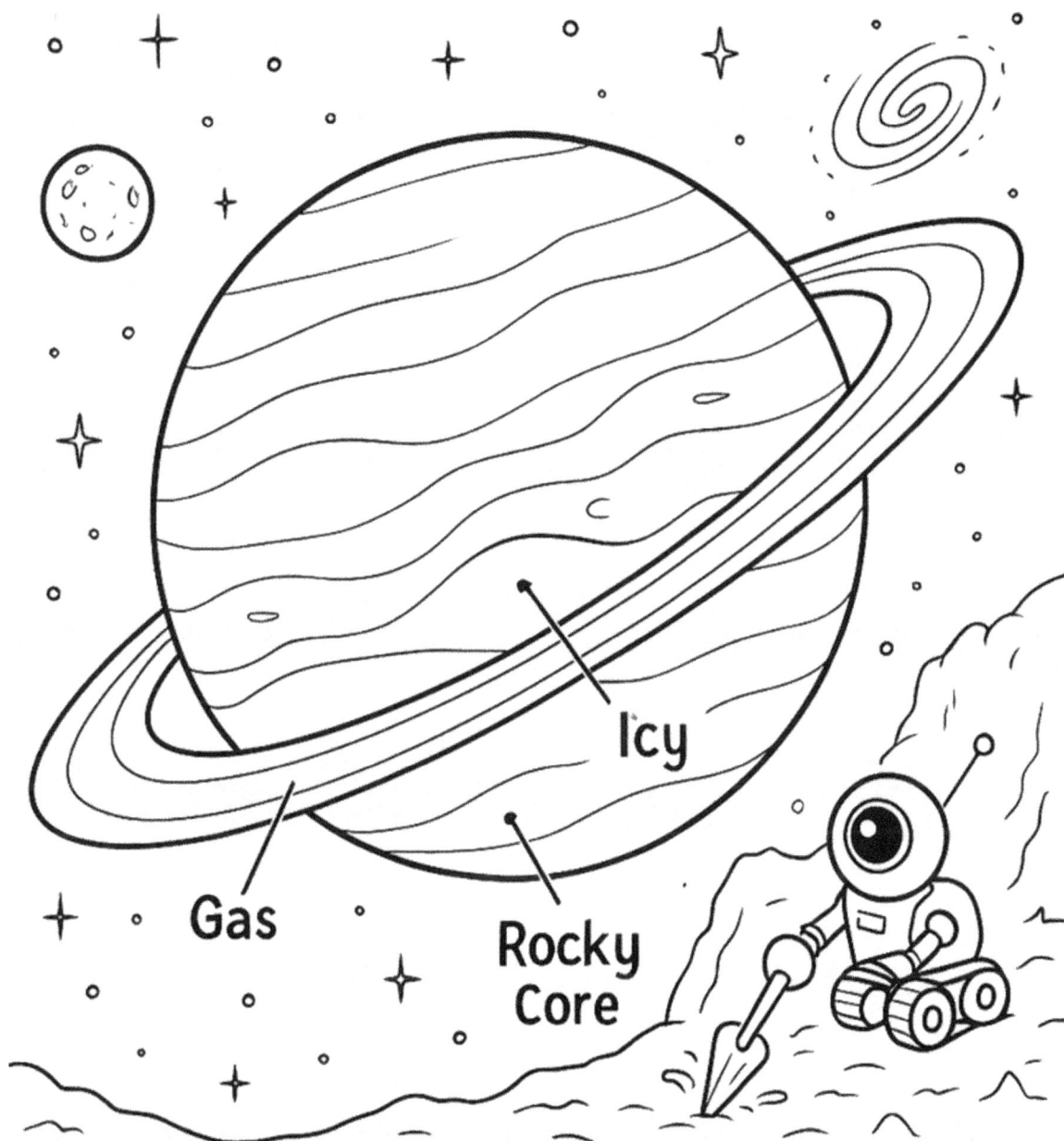

Icy

Gas

Rocky
Core

Uranus is called an ice giant – it has a thick layer of icy fluids!

Its core might be small, rocky, and super hot!

Uranus is made of gas on the outside, ice in the middle, and rock deep inside!

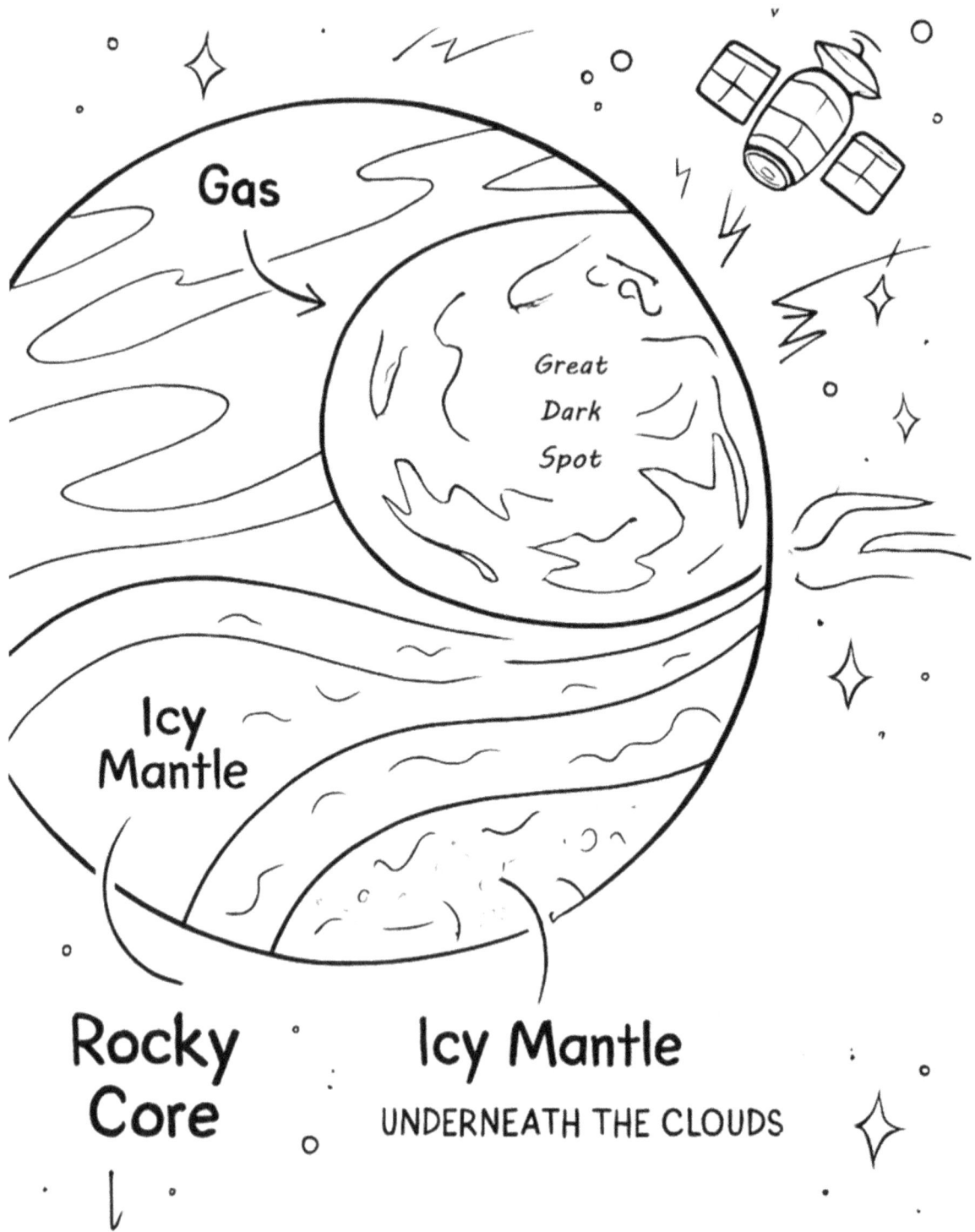

NEPTUNE!

Gas

Great
Dark
Spot

Icy
Mantle

Rocky
Core

Icy Mantle
UNDERNEATH THE CLOUDS

INSIDE NEPTUNE

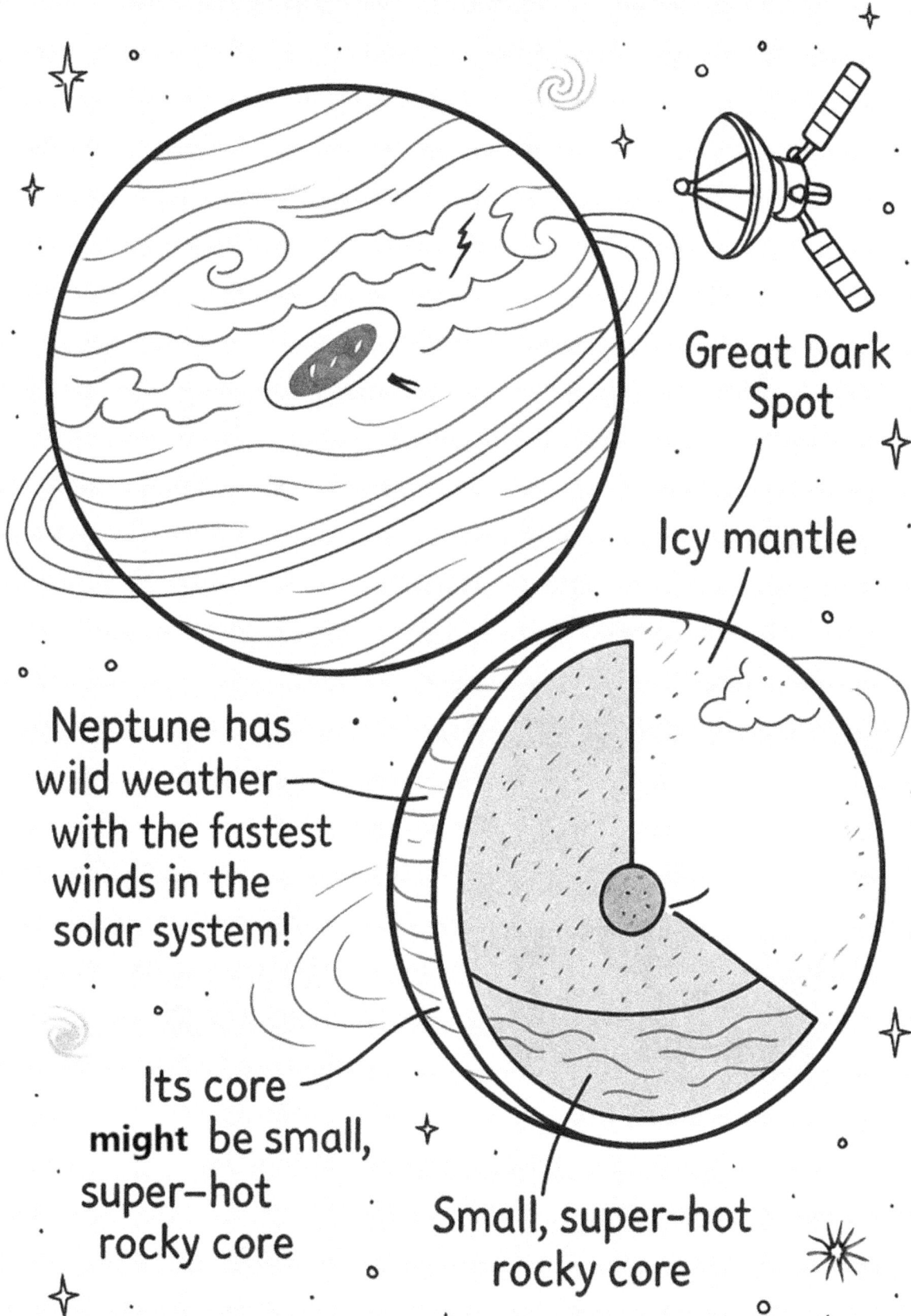

Great Dark Spot

Icy mantle

Neptune has wild weather with the fastest winds in the solar system!

Its core **might** be small, super-hot rocky core

Small, super-hot rocky core

PLUTO

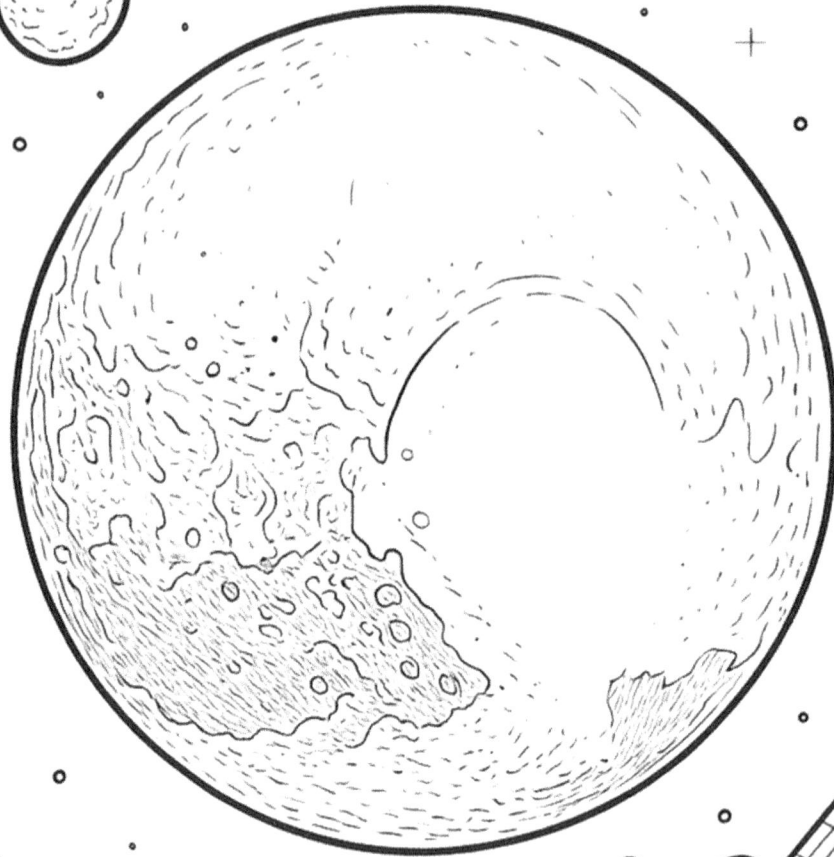

Pluto is a tiny world
of ice and rock far
beyond Neptune!

Its heart-shaped region is called
Tombaugh Regio!

THE SUN

The Sun is a giant ball of hot gas — a star!

It gives light and heat to all the planets!

It's 93 million miles from Earth

INSIDE THE SUN!

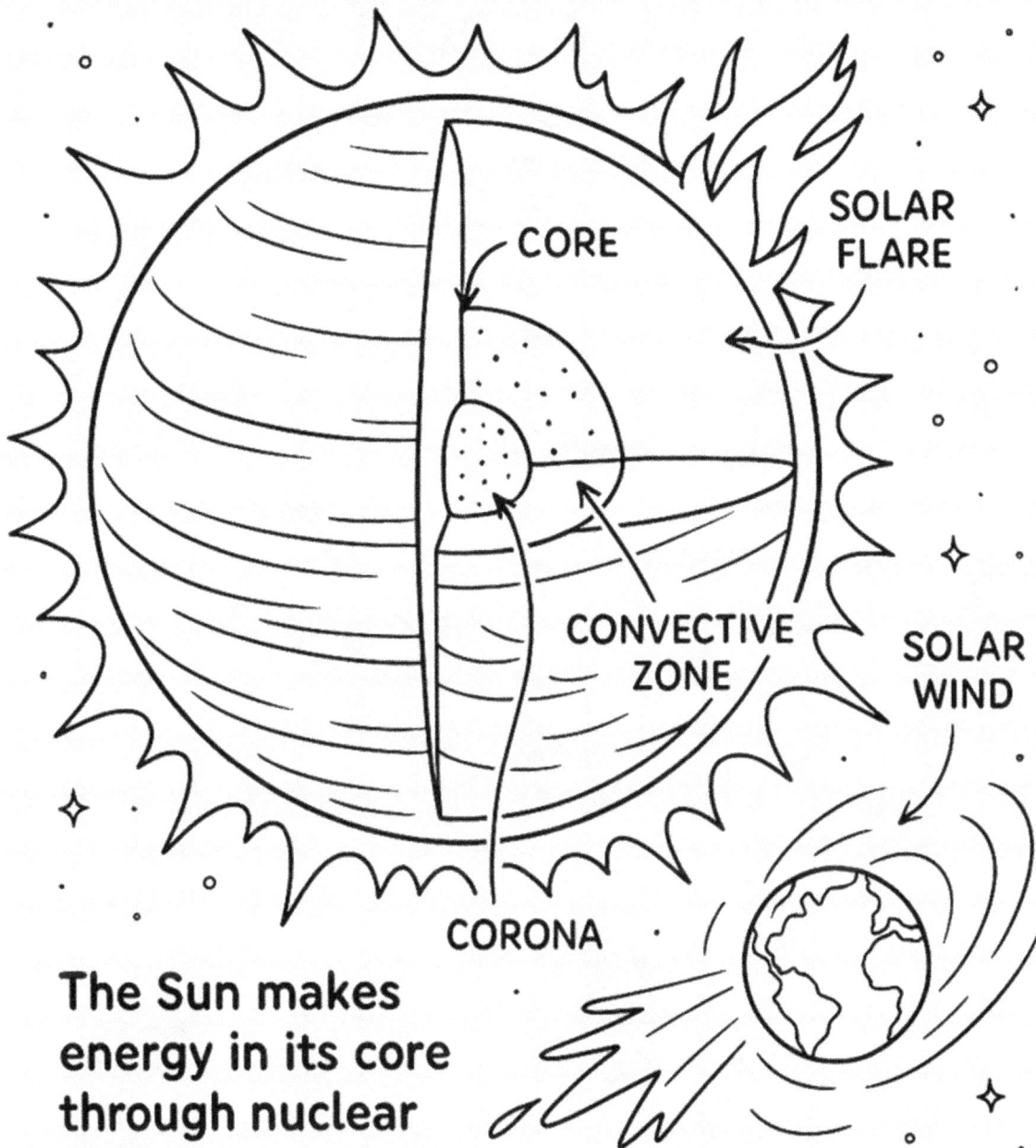

CORE

SOLAR FLARE

CONVECTIVE ZONE

SOLAR WIND

CORONA

The Sun makes energy in its core through nuclear fusion!

That energy travels through layers to reach the surface.

Its outer atmosphere, the corona, shoots out solar wind!

BLACK HOLES: COSMIC VACUUMS!

Spaghettification is real!

Nothing can escape a black hole — not even light!

Black holes are invisible— we see them by how tiey affect nearby stuff!

ORION NEBULA

The Orion Nebula is a place where new stars are born!

It's made of glowing gas and dust—like a cosmic cloud!

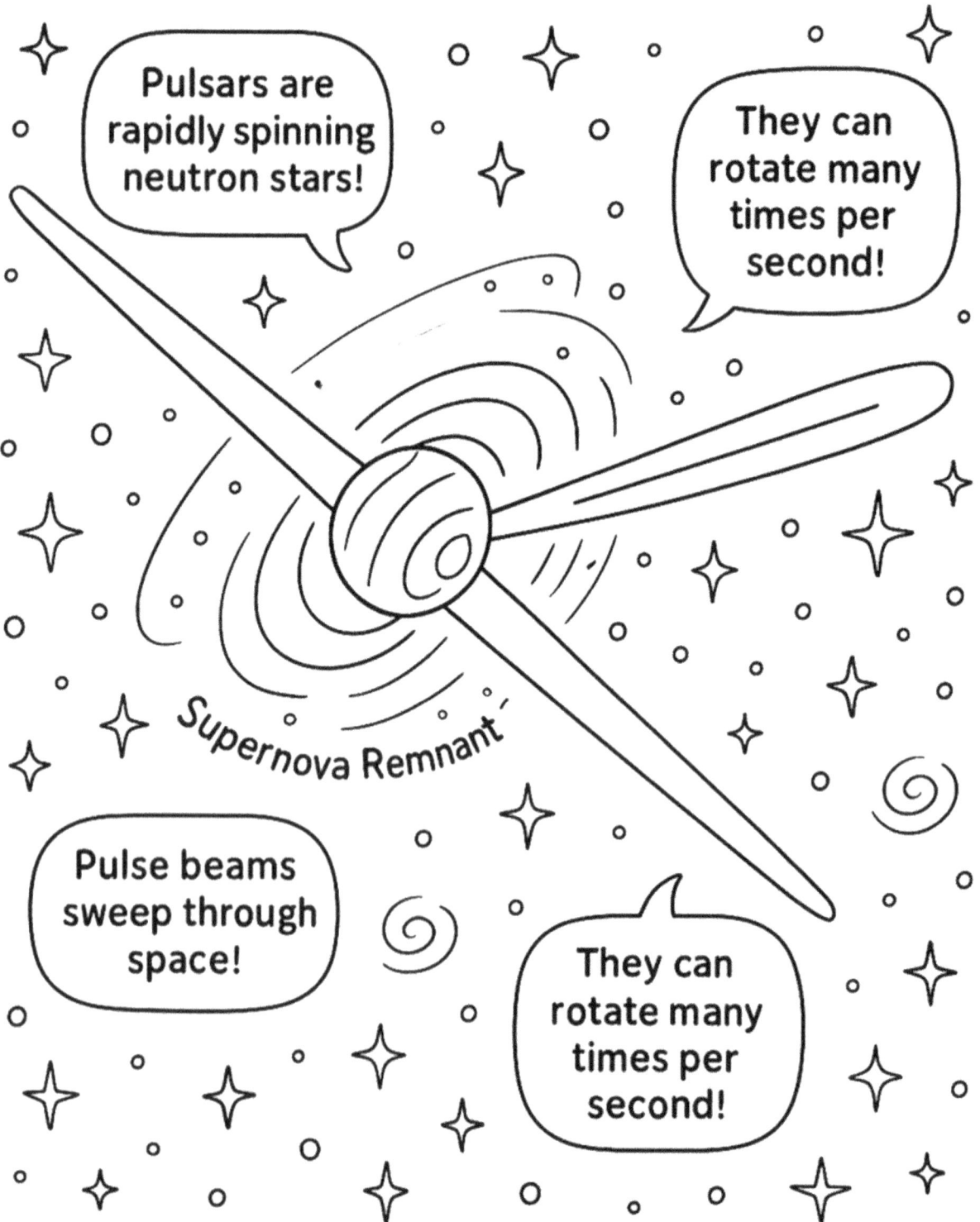

CONNECT THE STARS:
ORION!

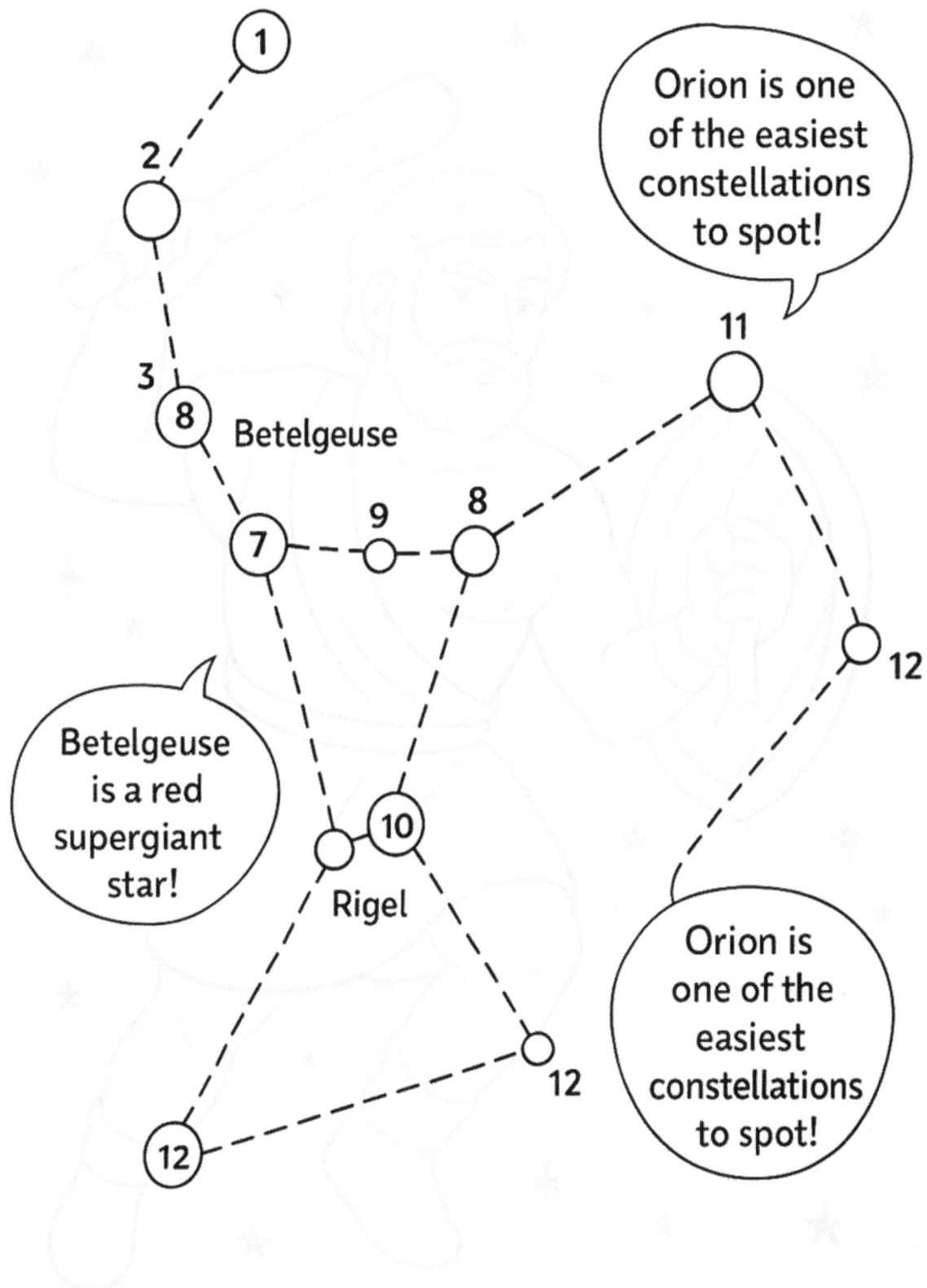

CONNECT THE STARS: THE BIG & LITTLE DIPPER!

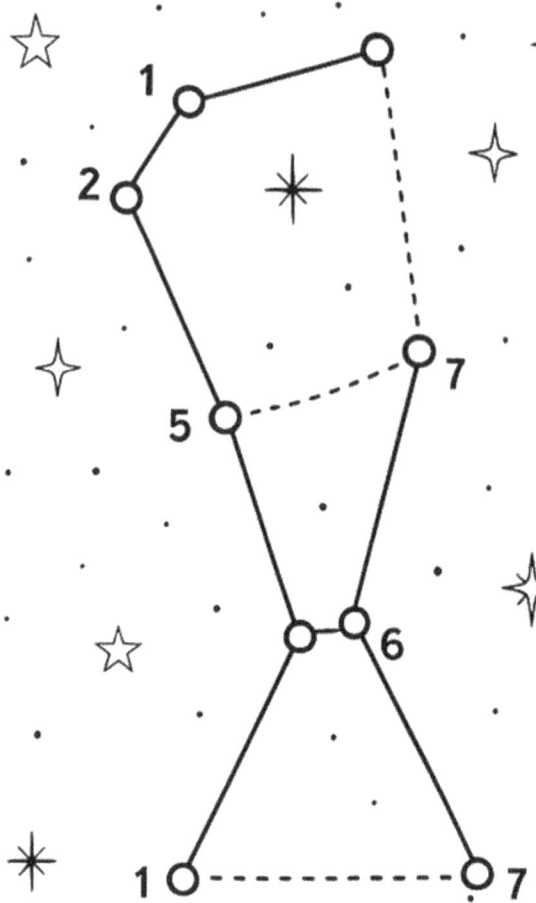

Polaris

The Big Dipper points to Polaris!

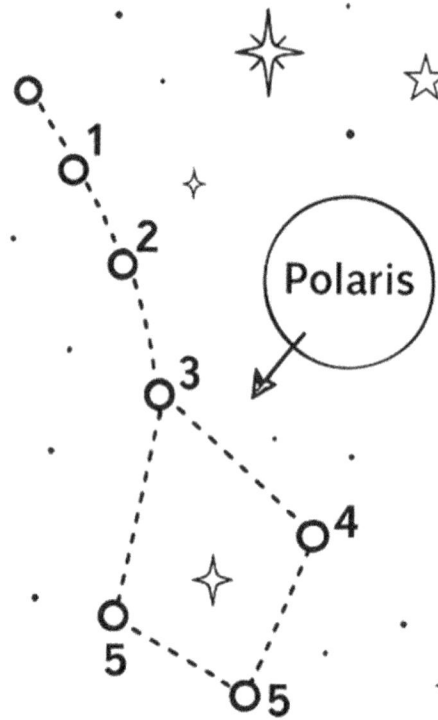

You can see these dippers year-round in the northern sky

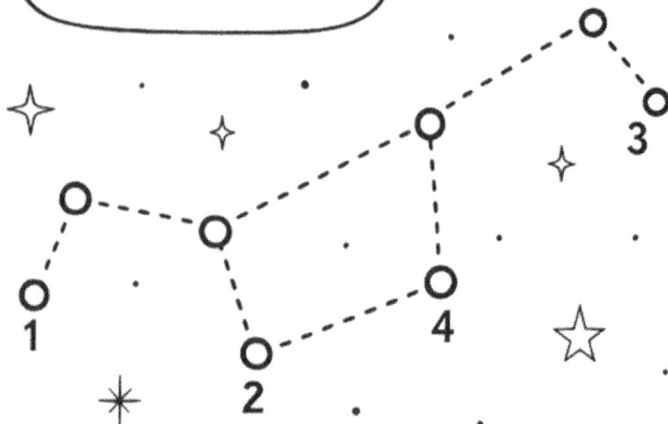

You can see these dippers year-round in the northern sky

PLANET MATCH-UP: WHICH PLANET AM I?

Draw a line from each planet to its fun fact!

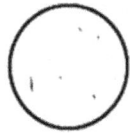 Mercury

 Venus

 Earth

 Mars

 Jupiter

 Saturn

 Uranus

 Neptune

- I'm the hottest planet in the solar system!

- I'm the red one with dust storms.

- I'm the smallest and closest to the Sun.

- I'm the only planet known to support life.

- I have over 90 moons!

- I'm famous for my beautiful rings.

- I spin on my side!

- I'm dark blue and very windy.

Space Maze Adventure!

START

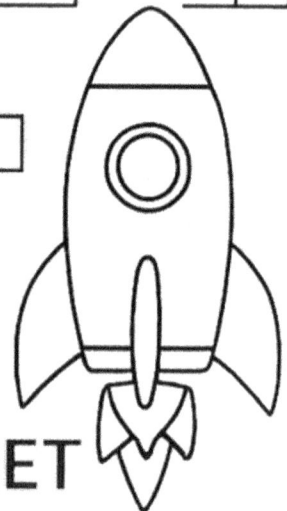

ROCKET

Design Your Own Spaceship!

Draw windows, wings, rocket flames--make it amazing!

Add
ROCKET
BOOSTERS

Design
contro!
pane!!

WHERE WILL
YOUR
SPACESHIP
GO?

DESIGN YOUR OWN PLANET!

No rules! Your planet can be anything!

Is it made of marshmallews? Has rainbow You decide?

Name of your planet: _____

What color is the sky? _____

What lives on this planet? _____

Does it have rings, oceans, or moon? _____

What is the weather like? _____

What makes this planet special? _____

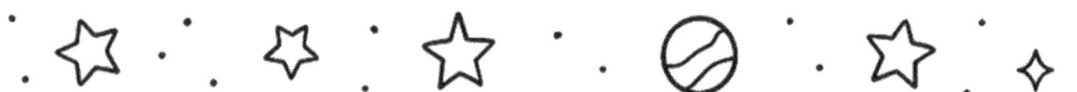

CERTIFICATE OF GALACTIC ACHIEVEMENT

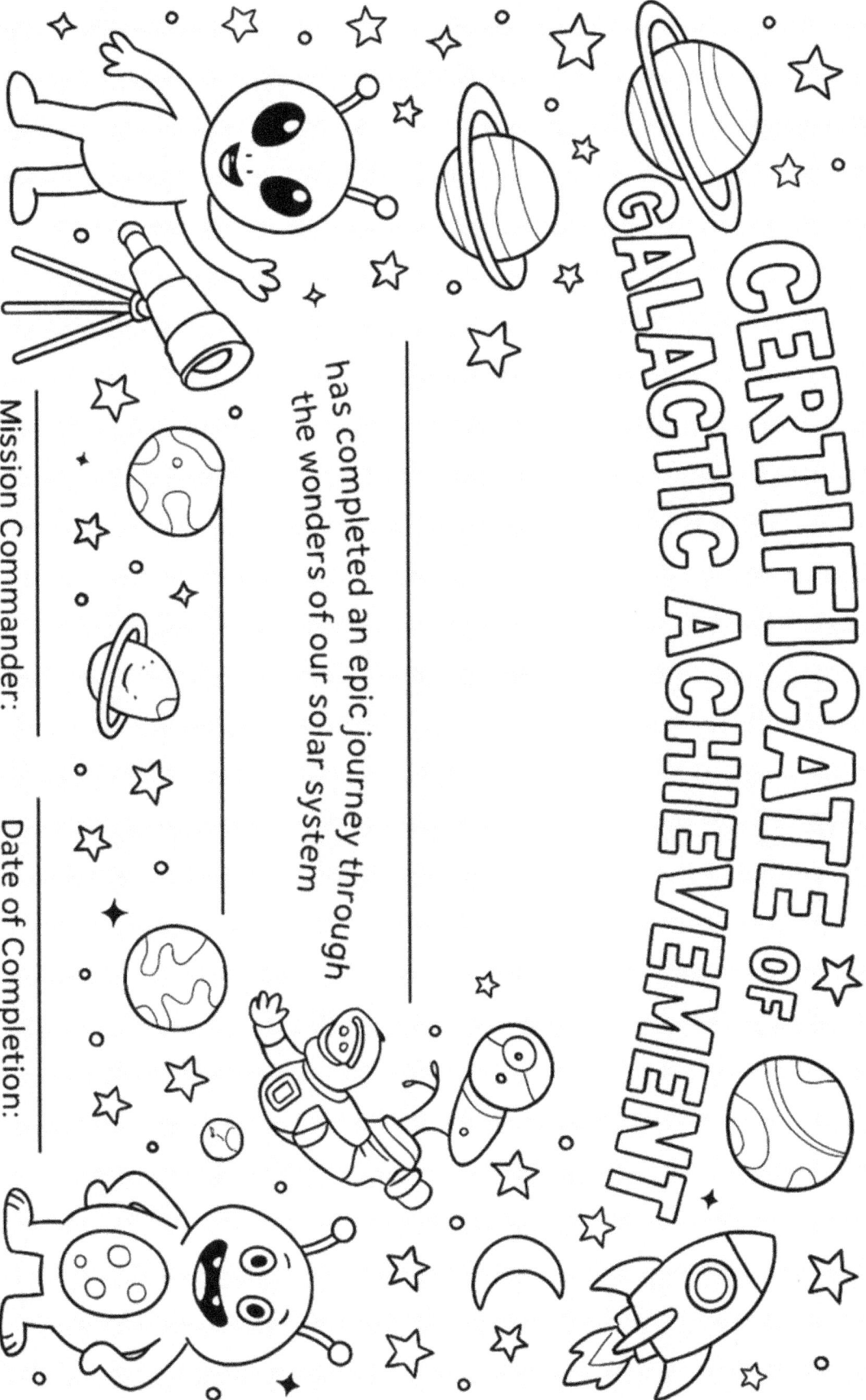

has completed an epic journey through the wonders of our solar system

Mission Commander:

Date of Completion: